WATCH ME PRACTICE

Math Workbook

Grade 1

Watch Me Practice Grade 1 Math Workbook

For more information about this title or to order other books and/or electronic media, contact the publisher:

B.L. Academic Services, LLC

http://watchmepracticeseries.com

info@watchmepracticeseries.com

First Edition

ISBN: 978-1-7369237-0-2

Printed in the United States

Cover Designer: Jamira Ink Designs LLC

Interior Designer/Editor: 1106 Design

Table of Contents

GRADE 1 SKILL REVIEW PRACTICE

Counting to Ten

Write 1–10 Using *Numbers*   **Write 1–10 Using *Words***

1. Write a number that is greater than 5. _____

2. Write a number that is less than 5. _____

3. What number comes after 7? _____

4. What number comes between 2 and 4? _____

5. What number comes next, counting backwards? 4, 3, 2, _____

Counting to Twenty

Write 1–20 Using *Numbers*

Write 1–20 Using *Words*

1. Write a number that is greater than 15. _____

2. Write a number that is less than 15. _____

3. What number comes after 12? _____

4. What number comes between 17 and 19? _____

5. What number comes next, counting backwards? 14, 13, _____ , 11

Counting by Twos

Counting by twos is called "skip counting," because every other number is skipped.

For example: 2, (skip 3), **4**, (skip 5), **6**, (skip 7), **8**

Skip counting by twos looks like this: 2, 4, 6, 8

Complete.

1. If you are skip counting by twos, what number would come after 8? _____

2. If you are skip counting by twos, what number would come after 12? _____

3. If you are skip counting by twos, what number would come after 16? _____

Counting by Fives

Counting by fives is also called "skip counting," because every fourth number is skipped.

For example:

5, (skip 6, skip 7, skip 8, skip 9) **10**, (skip 11, skip 12, skip 13, skip 14) **15**

Skip counting by fives looks like this: 5, 10, 15

Complete.

1. If you are skip counting by fives, what number would come after 15? _____

Addition

Addition adds to the amount you have. **For example:**

If you have 2 grapes and your mom gives you 2 more grapes, how many grapes do you now have? _____. Do you now have more grapes or fewer grapes? _____
Good Job!

Addition Facts

Knowing your addition facts will help you add numbers. Practice these addition facts. Fill in the blanks.

Adding 0	Adding 1	Adding 2	Adding 3	Adding 4
$0 + 0 = 0$	$1 + 0 = 1$	$2 + 0 = 2$	$3 + 0 =$	$4 + 0 = 4$
$0 + 1 = 1$	$1 + 1 = 2$	$2 + 1 =$	$3 + 1 = 4$	$4 + 1 =$
$0 + 2 = 2$	$1 + 2 =$	$2 + 2 = 4$	$3 + 2 =$	$4 + 2 = 6$
$0 + 3 = 3$	$1 + 3 = 4$	$2 + 3 =$	$3 + 3 = 6$	$4 + 3 =$
$0 + 4 =$	$1 + 4 =$	$2 + 4 = 6$	$3 + 4 =$	$4 + 4 = 8$
$0 + 5 =$	$1 + 5 = 6$	$2 + 5 =$	$3 + 5 = 8$	$4 + 5 =$
$0 + 6 = 6$	$1 + 6 =$	$2 + 6 = 8$	$3 + 6 =$	$4 + 6 = 10$
$0 + 7 =$	$1 + 7 = 8$	$2 + 7 =$	$3 + 7 = 10$	$4 + 7 =$
$0 + 8 = 8$	$1 + 8 =$	$2 + 8 = 10$	$3 + 8 =$	$4 + 8 = 12$
$0 + 9 = 9$	$1 + 9 = 10$	$2 + 9 =$	$3 + 9 = 12$	$4 + 9 =$

1. $1 + 1 =$ _____
2. $2 + 8 =$ _____
3. $3 + 9 =$ _____
4. $4 + 6 =$ _____
5. $0 + 7 =$ _____

6. $1 + 5 =$ _____
7. $2 + 4 =$ _____
8. $3 + 3 =$ _____
9. $4 + 9 =$ _____
10. $1 + 7 =$ _____

Subtraction

Subtraction takes away from the amount you have. **For example:**

If you had 4 cookies but gave away 2 cookies, how many cookies would you have left? _____. Would you have more cookies left or fewer cookies left? _____.

Good Job!

Subtraction Facts

Knowing your subtraction facts will help you subtract numbers. You cannot subtract a larger number from a smaller number.

For example: $1 - 2 =$ You cannot subtract 2 from 1.

Practice these subtraction facts. Fill in the blanks.

Subtracting 0	Subtracting 1	Subtracting 2	Subtracting 3	Subtracting 4
$0 - 0 = 0$	$1 - 1 = 0$	$2 - 2 = 0$	$3 - 3 =$	$4 - 4 = 0$
$1 - 0 = 1$	$2 - 1 = 1$	$3 - 2 =$	$4 - 3 = 1$	$5 - 4 =$
$2 - 0 = 2$	$3 - 1 =$	$4 - 2 = 2$	$5 - 3 =$	$6 - 4 = 2$
$3 - 0 =$	$4 - 1 = 3$	$5 - 2 =$	$6 - 3 = 3$	$7 - 4 =$
$4 - 0 = 4$	$5 - 1 = 4$	$6 - 2 = 4$	$7 - 3 =$	$8 - 4 = 4$
$5 - 0 = 5$	$6 - 1 =$	$7 - 2 =$	$8 - 3 = 5$	$9 - 4 =$
$6 - 0 =$	$7 - 1 = 6$	$8 - 2 = 6$	$9 - 3 =$	
$7 - 0 = 7$	$8 - 1 = 7$	$9 - 2 =$		
$8 - 0 = 8$	$9 - 1 =$			
$9 - 0 =$				

Complete:

1. 9 – 0 = _____

2. 6 – 1 = _____

3. 7 – 2 = _____

4. 5 – 3 = _____

5. 8 – 4 = _____

6. 0 – 0 = _____

7. 3 – 1 = _____

8. 2 – 2 = _____

9. 9 – 3 = _____

10. 6 – 4 = _____

Calendar Skills

1. How many months are in a year? _____

2. What month were you born? _____

3. SAY the names of the months.

4. How many seasons are in a year? _____

5. SAY the names of the seasons.

6. How many days are there in a week? _____

7. SAY the names of the days of the week.

Counting to 50

Practice counting to 50 using numbers.

1	2	3	4	5
6	7	8	9	10
11				

Practice counting to 50 using words.

One	Two	Three	Four	Five
Six	Seven	Eight	Nine	Ten
Eleven				

Counting by Twos

You learned in kindergarten that counting by twos is also called "skip counting," because every other number is skipped. **For example:**

2, (skip 3) **4**, (skip 5), **6**, (skip 7), **8**, (skip 9) **10**

Skip counting by twos looks like this: 2, 4, 6, 8, 10 . . . .

Complete:

Counting by twos, complete the following chart:

10		14		18
20	22		26	
30		34		38
40	42		46	
50				

Counting by Fives

Counting by fives is also called "skip counting," because every fourth number is skipped. **For example:**

5, (skip 6, 7, 8, 9) **10**, (skip 11, 12, 13, 14) **15**, (skip 16, 17, 18, 19) **20**

Skip counting by fives looks like this: 5, 10, 15, 20 . . . .

Complete:

Counting by fives, complete the following chart:

5	10	15	20	
30		40		50

Counting by Tens

Counting by tens is also called "skip counting," because every ninth number is skipped. **For example:**

10, (skip 11, 12, 13, 14, 15, 16, 17, 18, 19) **20**

Skip counting by tens looks like this: 10, 20, 30 . . .

Counting by tens, complete the following chart:

10	20			

Skip Counting by Twos, Fives, and Tens

1. If you are skip counting by twos, what number would come after 12? _____
2. If you are skip counting by fives, what number would come after 25? _____
3. If you are skip counting by tens, what number would come after 20? _____
4. If you are skip counting by twos, what number would come after 24? _____
5. If you are skip counting by fives, what number would come after 45? _____
6. If you are skip counting by tens, what number would come after 40? _____
7. If you are skip counting by twos, what number would come after 17? _____
8. If you are skip counting by fives, what number would come after 22? _____
9. If you are skip counting by tens, what number would come after 11? _____
10. If you are skip counting by twos, what number would come after 33? _____

Place Value

Digits in a number have a place value.

For example: The number 45 has a tens place and ones place. The 4 is in the tens place, and the 5 is in the ones place.

$$4 \text{ tens} = 10 + 10 + 10 + 10 = 40$$
$$5 \text{ ones} = 1 + 1 + 1 + 1 + 1 = 5$$
$$45$$

Identify the place value of each digit:

Number	Tens	Ones
37	3	7
12		
29		
46		

Greater Than, Less Than, Equal To

The > greater than sign opens to the larger number.

The < less than sign points to the lesser number.

The = equal sign says both numbers are the same.

Use the >, <, or = signs to complete the following:

1. 2 is _____ 3

2. 10 is _____ 5

3. 13 is _____ 13

4. 25 is _____ 50

5. 47 is _____ 37

What Number Comes Before, After, or Between?

1. What number comes before 40? _____

2. What number comes after 32? _____

3. Does number 65 come before or after 66? _____

4. What number comes between 35 and 37? _____

5. What number comes between 28 and 30? _____

6. What number comes after 18? _____

Ordinal Numbers

Ordinal numbers tell the position of numbers.

For example: If there are 10 people in a line, the person at the beginning of the line is called the first person. The last person in line is called the tenth — or 10th — person in line.

Practice learning ordinal numbers 1–10.

First	**Sixth**
Second	**Seventh**
Third	**Eighth**
Fourth	**Ninth**
Fifth	**Tenth**

Here is a line of 10 circles:

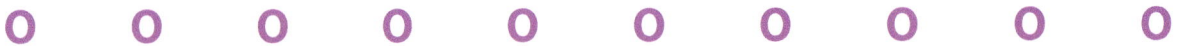

O   O   O   O   O   O   O   O   O   O

1. Point to the 3rd circle.

2. Point to the 5th circle.

3. Point to the 8th circle.

4. Point to the 6th circle.

ADDITION

Addition increases the amount of something. When you add numbers together, the answer you get is called the sum.

For example:

- If you had 3 cookies and your teacher gave you 2 more cookies, how many cookies would you have?   <u>5</u>

- How many cookies did you have before you were given the 2 cookies?

  _____

- Do you have more cookies after getting the 2 cookies or fewer cookies?

  _____

The number sentence for this example is 3 + 2 = 5

Knowing your addition facts will help you add numbers. In kindergarten, you learned to add with numbers 0 through 4. Practice adding with numbers 5 through 12. **Fill in the blanks.**

Adding 5	Adding 6	Adding 7	Adding 8	Adding 9
$5 + 0 = 5$	$6 + 0 =$	$7 + 0 = 7$	$8 + 0 =$	$9 + 0 = 9$
$5 + 1 =$	$6 + 1 = 7$	$7 + 1 =$	$8 + 1 = 9$	$9 + 1 =$
$5 + 2 = 7$	$6 + 2 = 8$	$7 + 2 = 9$	$8 + 2 = 10$	$9 + 2 = 11$
$5 + 3 = 8$	$6 + 3 = 9$	$7 + 3 = 10$	$8 + 3 =$	$9 + 3 = 12$
$5 + 4 = 9$	$6 + 4 =$	$7 + 4 = 11$	$8 + 4 = 12$	$9 + 4 =$
$5 + 5 =$	$6 + 5 = 11$	$7 + 5 =$	$8 + 5 = 13$	$9 + 5 = 14$
$5 + 6 = 11$	$6 + 6 = 12$	$7 + 6 = 13$	$8 + 6 =$	$9 + 6 = 15$
$5 + 7 = 12$	$6 + 7 = 13$	$7 + 7 = 14$	$8 + 7 = 15$	$9 + 7 =$
$5 + 8 =$	$6 + 8 =$	$7 + 8 =$	$8 + 8 = 16$	$9 + 8 = 17$
$5 + 9 = 14$	$6 + 9 = 15$	$7 + 9 = 16$	$8 + 9 =$	$9 + 9 = 18$
$5 + 10 =$	$6 + 10 = 16$	$7 + 10 =$	$8 + 10 = 18$	$9 + 10 =$

Add.

1. 9
   $+ 7$

2. 6
   $+ 2$

3. 8
   $+ 5$

4. 7
   $+ 4$

5. 9
   $+ 3$

Number Sentences

$5 + 4 =$ _____

$8 + 2 =$ _____

Adding 10	Adding 11	Adding 12
10 + 0 = 10	11 + 0 =	12 + 0 = 12
10 + 1 = 11	11 + 1 = 12	12 + 1 = 13
10 + 2 = 12	11 + 2 = 13	12 + 2 = 14
10 + 3 = 13	11 + 3 = 14	12 + 3 =
10 + 4 =	11 + 4 =	12 + 4 = 16
10 + 5 = 15	11 + 5 = 16	12 + 5 = 17
10 + 6 = 16	11 + 6 = 17	12 + 6 =
10 + 7 = 17	11 + 7 = 18	12 + 7 = 19
10 + 8 =	11 + 8 =	12 + 8 = 20
10 + 9 = 19	11 + 9 = 20	12 + 9 =
10 + 10 = 20	11 + 10 =	12 + 10 = 22

Addition Word Problems

1. Emily had a red, a blue, and a green crayon. She borrowed a black, a brown, and a pink crayon from Madison. How many crayons does Emily now have?

   _____

2. Write a number sentence to show the sum of Emily's crayons.

   _____

3. Tyler has six marbles. He needs ten marbles to play his game. How many more marbles does Tyler need to make ten? _____
   6 + _____ = 10.

SUBTRACTION

Subtraction decreases the amount of something. The answer you get when you subtract numbers is called the **difference**.

For example:

- If you had 5 lollipops but gave 3 of the lollipops away, how many lollipops would you have left?   2

- How many lollipops did you have before giving away 3 lollipops?   _____

- Did you have more lollipops left after giving away 3 lollipops or fewer lollipops?

  _____

The number sentence for this example is $5 - 3 = 2$

Knowing your subtraction facts will help you subtract numbers. In kindergarten you learned to subtract with numbers 0 through 4. Practice subtracting with numbers 5 through 12.

Remember: You cannot subtract a larger number from a smaller number.

For example: $5 - 6 =$ You cannot subtract 6 from 5

Practice these subtraction facts. **Fill in the blanks.**

Subtracting 5	Subtracting 6	Subtracting 7	Subtracting 8	Subtracting 9
$5 - 5 = 0$	$6 - 6 =$	$7 - 7 = 0$	$8 - 8 =$	$9 - 9 = 0$
$6 - 5 = 1$	$7 - 6 = 1$	$8 - 7 = 1$	$9 - 8 = 1$	$10 - 9 =$
$7 - 5 =$	$8 - 6 =$	$9 - 7 =$	$10 - 8 =$	
$8 - 5 = 3$	$9 - 6 = 3$	$10 - 7 =$		
$9 - 5 = 4$	$10 - 6 =$			
$10 - 5 =$				

Subtracting 10	Subtracting 11	Subtracting 12
10 – 10 =	11 – 11 = 0	12 – 12 =
11 – 10 =	12 – 11 =	
12 – 10 = 2		

Subtract.

1. 10
  - 6

2. 9
  - 4

3. 7
  - 5

4. 8
  - 8

5. 11
  - 1

Number Sentences

- 12 – 12 = _____

- 7 – 4 = _____

Subtraction Word Problems

1. Ashley had eight balloons. Kelly popped three of the balloons. How many balloons does Ashley have left? _____

2. Write a number sentence to show how many balloons are left.

   _____

3. For dinner, Angelica had eight chicken nuggets on her plate. Her dog, Chiloh, jumped on the table and ate four of the nuggets. How many chicken nuggets does Angelica have left? _____

   8 – 4 = _____

How Much Is This Coin Worth?

1. What is a penny worth? _____

2. How much is a nickel worth? _____

3. How much is a dime worth? _____

4. How much is a quarter worth? _____

5. What coins would you need to show 12 cents? _____

6. What coins would you need to show 15 cents? _____

7. What two coins would you need to show 26 cents? _____

8. What coin is worth the same as two dimes and a nickel? _____

9. If I gave you a dime, what coins could you give me that would equal the same amount?

   _____

10. Tara had three nickels, and Kelley had a nickel and a dime. Who had more money?

    _____

    Explain your answer. _____

    _____

TELLING TIME

The short hand on the face of a clock is called the **hour hand**. The hour hand tells the hour.

The long hand on the face of a clock is called the **minute hand**. The minute hand tells the time to the minute.

For example: If the hour hand is on the 3 and the minute hand is on the 12, the time is 3 o'clock or 3:00.

1. If the hour hand is on the 4 and the minute hand is on the 12, what time would it be?

   _____

2. If the hour hand is on the 12 and the minute hand is on the 12, what time would it be?

   _____

3. If the hour hand is on the 6 and the minute hand is on the 12, what time would it be?

   _____

USING A CALENDAR

A calendar tells the months, days of the week, number of days, and the year.

1. How many months are in one year? _____

2. How many days are in a week? _____

3. Write the days of the week on the lines below:

_____

_____

_____

_____

_____

_____

_____

4. How many days are in the month of June? _____

5. How many days are in January? _____

6. What month is Christmas? _____

7. What month is Halloween? _____

8. What two days are called "the weekend"?

_____

_____

This graph shows the desserts that students at the Sunny Side Elementary School enjoy eating.

Desserts	5	10	15	20
Cookies	■	■	■	
Cake	■	■		
Pie	■			
Ice cream	■	■	■	■

1. Which dessert is the most favorite? _____
   How do you know? _____
   _____
   _____

2. Which dessert is the least favorite? _____
   How do you know? _____
   _____
   _____

3. Which dessert is the 2nd most favorite? _____
   How do you know? _____
   _____
   _____

4. Which dessert is the 3rd most favorite? _____
   How do you know? _____
   _____
   _____

FRACTIONS

A fraction has a top number and a bottom number. The bottom number tells the entire amount. The top number tells the part that is being used. The top number is called the numerator. The bottom number is called the denominator.

For example:
$$\frac{3}{4}$$ numerator
denominator

This fraction tells us there are 4 parts to this fraction but only 3 parts are being used.

To write fractions, first find the entire amount. Then find the amount that is being used.

For example:
$$\frac{5}{6}$$

a. What is the entire amount of this fraction? _____

b. What is the amount that is being used? _____

1. Write a fraction with a numerator of 4 and a denominator of 6. _____

_____

2. Write a fraction with a numerator of 7 and a denominator of 8. _____

_____

3. Draw a circle. Color half of the circle. Write a fraction that represents your drawing.

_____